The Fundamental Spin Force

The Origin of Attraction and Repulsion

Fundamental Physics

Helmut M. Albert

The Fundamental Spin Force

The Origin of Attraction and Repulsion

Fundamental Physics

Helmut M. Albert

Imprint

Bibliografische Information der Deutschen Nationalbibliothek: Die Deutsche Nationalbibliothek verzeichnet diese Publikation in der Deutschen Nationalbibliografie; detaillierte bibliografische Daten sind im Internet über dnb.dnb.de abrufbar.

Bibliographic information from the German National Library: The German National Library lists this publication in the German National Bibliography; detailed bibliographic data is available online at dnb.dnb.de.

Publisher: BoD · Books on Demand GmbH, Überseering 33, 22297 Hamburg, bod@bod.de
Print: Libri Plureos GmbH, Friedensallee 273, 22763 Hamburg

ISBN: 978-3-6963-7620-8

Content

Preface

The question of the fundamental cohesion of the universe has occupied scientists for centuries. Modern physics has developed sophisticated models to describe interactions; however, these approaches primarily focus on the description of observable effects, while certain conceptual aspects—particularly the interpretation of spin—remain an open question.

My engagement with these questions arose from an interdisciplinary perspective that combines artistic work with an interest in physical structures and forms. This approach led me to consider whether fundamental phenomena might be understood in a more direct, mechanical way.

An important foundation for this work is provided by earlier manuscripts on atomic structure, which my father developed as early as the 1960s. Under the title *"Atomic Model with Checkerboard Structure"*, I published these manuscripts ten years ago. In the present book, spin—described in physics as an abstract property—is considered as a real physical rotation. Based on this interpretation, it is proposed that attraction and repulsion result from rotational states and that fundamental interactions can be described within a unified framework.

The present English edition is not merely a translation of the original German text, but also a revised and refined version. In the course of translation, various terms were specified more precisely, the terminology was standardized, and certain theoretical aspects were further developed and deepened.

The English text was created with the assistance of AI-based language tools. However, all underlying ideas, concepts, and theoretical considerations originate from the author.

The reader is invited to approach the arguments presented here with an open and critical mind.

Helmut M. Albert

Freiburg, April 2026

Chapter 1 - Spin and Force

In physics, the fundamental forces of nature are described in terms of four interactions. Essential insights into these forces and the structure of the universe were already developed between the 17th and 19th centuries and laid the foundation for modern science. The laws formulated by Isaac Newton and Charles-Augustin de Coulomb illustrate the duality of forces and their effects. Building on Newtonian theory, Immanuel Kant described attraction and repulsion as fundamental forces of the universe in 1755, thereby offering an early explanation of celestial mechanics (Kant 1755; 1925, p. 19).

At that time, however, the microscopic world of rotating particles was still unknown. It was not until the 20th century that the fundamental constituents of matter and their property of spin were discovered. The concept of particle spin, together with other discoveries, led to the development of new theories that marked the beginning of modern physics, or quantum physics. The simple and coherent laws of nature established in earlier centuries were thus significantly expanded. At the same time, however, these developments also introduced interpretations that remain subject to ongoing discussion.

While the theories of Newton and Coulomb required only two interacting bodies to describe forces, 20th-century physics introduced an additional element: force-carrying or exchange particles. These particles, known as bosons, are assumed to mediate interactions between matter particles and to account for the phenomena of attraction and repulsion.

However, the physical mechanism underlying certain proposed exchange particles—such as gluons or gravitons—cannot be directly described in mechanical terms. This raises fundamental questions: how interactions can be understood at a deeper level, which forces are actually at work, and what their underlying source of energy is?

There is one phenomenon that can be observed and experimentally verified in all physically relevant particles, bodies, and systems: spin, that is, rotational motion. This observation leads to the central question of which properties of spin could account for the phenomena of attraction and repulsion. To address this question, the following chapters first present the established interaction theories and exchange-particle concepts according to the current state of research. The next chapter is devoted to the scientific understanding of spin and the history of its discovery. It reviews the results of international research on particle spin and highlights its significance in contemporary physics.

The final chapter then develops an alternative approach based on a mechanical interpretation of spin and a unified description of interactions. This approach builds on the concept of a checkerboard-like planar atomic structure (Albert 2023) and is intended to critically examine the prevailing theoretical framework.

Chapter 2 - Interactions

The Standard Model of elementary particle physics describes fundamental interactions—with the exception of gravitation—and distinguishes between matter particles and exchange particles, referred to as fermions and bosons (or gauge bosons). According to this model, bosons are considered "virtual" particles, also known as exchange particles, force carriers, or messenger particles. These particles are assumed to move back and forth between matter particles in rapid succession, thereby mediating fundamental forces.

In quantum field theory, many exchange particles are described as virtual and are not directly observable. This introduces a conceptual tension between empirical observability and theoretical description, motivating alternative interpretations of fundamental interactions, for example in terms of spin and rotational dynamics.

In this framework, bosons are assigned integer spin values (0, 1, 2), while fermions—including protons and neutrons—have a spin of ½. This distinction implies that fermions possess intrinsic spin, whereas bosons are assigned integer spin values and differ in their physical role from fermions. However, it remains a matter of debate whether this spin corresponds to a physically real rotation or should be understood purely as a quantum-mechanical property.
It is generally assumed that the forces observable in everyday life originate from interactions at the microscopic level, where exchange particles mediate

forces between elementary constituents (Roth and Stahl 2020, pp. 1950–1956). According to the University of Wuppertal (2023a, 2023b), these interaction theories have been extensively confirmed by experimental results over the past decades. This confirmation applies both to the existence of the particles themselves and to the interactions described within the Standard Model.

Nevertheless, certain limitations of this framework have been discussed. Demtröder (2014) notes that quantum chromodynamics (QCD), developed in analogy to quantum electrodynamics, represents an important theoretical approach but does not provide a fully convincing explanation in all respects. QCD is based on the quark model, which postulates that nucleons consist of quarks interacting through the exchange of gluons and other gauge bosons. Despite conceptual challenges, the model allows for quantitatively accurate predictions (Demtröder 2014, p. 32).

Bleck-Neuhaus describes the development of this theoretical framework as follows:

„Die Teilchenphysik erfand für die Wechselwirkung zwischen diesen eine neue Sorte Austauschteilchen, 'Gluon' genannt (Klebeteilchen'), mit Ladungen namens 'Farbe'. Als reelle Teilchen sind allerdings weder Gluonen noch Quarks bisher beobachtet worden"(Bleck-Neuhaus 2013, p. 8108).

> *("Particle physics introduced a new type of exchange particle for these interactions, called 'gluons' ('glue particles'), carrying charges known*

as 'color'. However, neither gluons nor quarks have so far been observed as real particles.")

This quotation highlights the conceptual nature of exchange particles within current theoretical models. Quantum chromodynamics explains the strong interaction as a particularly powerful force that binds quarks together. According to this theory, three quarks form a nucleon, and attempts to separate them result in the creation of new quark–antiquark pairs. The effective range of this interaction is limited to the scale of a nucleon, leading to the interpretation that the residual nuclear force between protons and neutrons is comparatively weak and mediated by virtual particle exchange (Bleck-Neuhaus 2013, pp. 8108–8119).

Meschede (2020) emphasizes that it was only through quantum chromodynamics that the quark model became physically comprehensible, in a way comparable to how electromagnetic interaction contributed to the understanding of atomic structure. In this context, quarks are considered carriers of color charge, while gluons mediate the interaction between them. Unlike electromagnetism, which involves two types of charge, the strong interaction involves three types of color charge.

Meschede summarizes this relationship as follows:

„Die starke Kraft oder Farbkraft hält die Quarks im Teilchen zusammen, aber nach einem ganz anderen Kraftgesetz als die Coulombkraft, die die Teilchen im Atom zusammenhält (Quark confinement)", (Meschede 2020, p. 983)

> *("The strong force or color force holds the quarks together within the particle, but according to a fundamentally different law than the Coulomb force that binds particles in the atom (quark confinement).")*

In contemporary nuclear physics, increasing emphasis is placed on precise experimental data. At facilities such as the Large Hadron Collider (LHC) at CERN, researchers conduct high-precision measurements to gain deeper insight into the nature of the strong interaction. According to current models, protons and neutrons consist of three quarks, while the interaction between nucleons is attributed to a residual force arising from quark interactions.

Recent experimental investigations have aimed to measure these forces more accurately. For example, results from the ALICE collaboration have been reported in *Nature* (Acharya et al. 2020; Goethe University Frankfurt 2020). In addition, other international research projects have examined the structure of atomic nuclei and observed unexpected behavior in the interaction between protons and neutrons.

In particular, findings suggest that under extreme conditions, the strong interaction may exhibit repulsive effects not only between protons but also between neutrons (Podbregar 2020; Schmidt et al. 2020). Such observations raise fundamental questions about the underlying mechanisms of interaction and suggest that intrinsic properties of particles—such as spin—may play a more direct role than previously assumed.

For this reason, the following chapter is devoted to a detailed examination of particle spin and its possible physical interpretation.

2.1 The Weak Interaction

Like the strong interaction, the weak interaction is assumed to act only over very short distances. Within the atomic nucleus, it is considered responsible for radioactive decay processes and for the transformation of neutrons into protons. The weak interaction also plays a crucial role in nuclear fusion in the Sun. In this process, protons are converted into neutrons, enabling the formation of helium nuclei.
Bleck-Neuhaus describes the weak interaction as follows: "Bei der elektromagnetischen Wechselwirkung wirken elektrisch geladene Teilchen mittels 'Photonen' aufeinander ein, während bei der schwachen Wechselwirkung 'W- und Z-Bosonen' ausgetauscht werden" (Bleck-Neuhaus 2014:8119–8121).

> *("In electromagnetic interaction, electrically charged particles interact via 'photons', whereas in the weak interaction 'W and Z bosons' are exchanged.")*

2.2 Strong Interaction

Historically, the strong interaction has also been referred to as the "strong force" or "strong nuclear force." Even before the formulation of modern theory in the 1970s, scientists had been investigating the forces acting between protons and neutrons, particularly following the discovery of the neutron. One of the first to propose a theoretical explanation was Hideki Yukawa.

In 1935, Yukawa postulated the existence of virtual exchange particles acting over short distances within the atomic nucleus. The particle he predicted was experimentally discovered in 1947 and later named the "pion" (Bleck-Neuhaus 2013: 8104).

With the development of higher-energy particle accelerators, further insights into nuclear interactions became possible. By the mid-1960s, scientists proposed that protons, neutrons, and mesons are composite particles consisting of more fundamental constituents, the so-called quarks.

Bleck-Neuhaus describes this development as follows: "Die Teilchenphysik erfand für die Wechselwirkung zwischen diesen eine neue Sorte Austauschteilchen, 'Gluon' genannt ('Klebeteilchen'), mit Ladungen namens 'Farbe'" (Bleck-Neuhaus 2013: 8104). ("Particle physics introduced a new type of exchange particle for these interactions, called 'gluons' ('glue particles'), carrying charges known as 'color'.")

He also emphasizes that these subparticles have not been directly observed. According to quantum chromodynamics (QCD), the strong interaction is so intense that quarks cannot be isolated. Instead, only bound states such as three quarks (forming nucleons) or quark–antiquark pairs (forming mesons) can exist (Bleck-Neuhaus 2013: 8104).

In general, atomic structure is understood in such a way that the strong interaction leads to an attraction between protons and neutrons that is significantly stronger than the electromagnetic interaction between electrons and the nucleus. However, as noted in the literature: „Die Beschreibung der Nukleonen ist jedoch ein offenes Problem". ("Quantenchromodynamik" 2023).

("The description of nucleons remains an open problem")

2.3 Electromagnetic Interaction

According to the theory of quantum electrodynamics (QED), photons act as exchange particles mediating interactions between charged particles and electromagnetic fields. As described by Meschede, the foundations of electromagnetic theory were established by Michael Faraday and James Clerk Maxwell, while the connection between electromagnetism and atomic structure was further developed by Ernest Rutherford and Niels Bohr. The theory of QED was later refined by Richard Feynman (Meschede 2020: 973).

A fundamental principle underlying electromagnetic interaction is Coulomb's law, formulated in the eighteenth century. It states that like charges repel each other, while opposite charges attract. This principle plays a central role in atomic structure: the Coulomb force acts attractively between protons and electrons and repulsively between protons.

This behavior contrasts with the interaction between protons and neutrons, which, according to conventional theory, is assumed to be exclusively attractive.

2.4 Gravitation

Gravitation differs from the other fundamental interactions in several respects. Most notably, it is not included in the Standard Model of particle physics. Furthermore, it remains unclear whether gravity can be described in terms of an exchange particle, the so-called "graviton."

On Earth, gravity manifests itself as a force directed toward the center of the planet. More generally, it is understood as the mutual attraction between masses and is considered responsible for the structure of planetary systems, solar systems, and galaxies.

Isaac Newton formulated the classical theory of gravitation, describing it as a force acting at a distance between two masses. According to this framework, all objects attract one another proportionally to their mass.

In the twentieth century, Albert Einstein fundamentally revised this concept. In his theory of general relativity, gravity is no longer interpreted as a force in the classical sense, but as a consequence of the curvature of spacetime.

Chapter 3 - The Particle Spin

In particle physics, fermions and bosons—considered the fundamental building blocks of matter—are distinguished by their spin. While fermions, which include protons and neutrons, have a spin of ½, bosons are assigned integer spin values (0, 1, or 2). This distinction is a central feature of modern quantum theory.

The discovery of the electron's spin, or intrinsic angular momentum, in 1925 occurred between the discoveries of the proton and the neutron. The two researchers Samuel Goudsmit and George Uhlenbeck identified this property through experimental and theoretical investigations. They recognized that electrons possess an intrinsic angular momentum, referred to as "spin."

However, this raises a fundamental conceptual difficulty: electrons appear to be point-like particles without spatial extension. This raises the question of how a particle without measurable size can be associated with rotation (Paetz 2020: 16–20).

An important experimental contribution to the understanding of spin was provided by the Stern–Gerlach experiment. This experiment demonstrated the quantization of angular momentum and showed that spin can assume discrete orientations (Paetz 2020: 22–23). Even today, spin measurements are closely related to principles established in this experiment (Camejo 2015: 225–227).

Although scientific interest in particle spin dates back to the early twentieth century, it has gained renewed importance in recent decades through the development of "spintronics." This field is based on the idea that spin, in addition to electric charge, can be used for information processing. As a result, researchers worldwide are working to exploit the spin of electrons and other particles for technological applications.

Determining the orientation of spin requires experimental measurement, which itself can influence the observed result. In order to measure the spin of protons, large ensembles of particles, such as protons or deuterons, are typically required. In their natural state, these spins are randomly oriented. By applying external fields, however, it is possible to align them in a common direction—a process known as polarization.

The internal structure of atoms can only be investigated indirectly through interactions between particles. Such investigations are commonly carried out in scattering experiments, in which particles collide with one another (Hejny 2016). When unpolarized particles are involved, the scattering distribution is typically symmetric. In contrast, polarized particles produce asymmetric distributions, allowing conclusions to be drawn about spin orientation.

To measure these effects, detectors are arranged around particle beams, often in a segmented, circular geometry. In high-energy experiments, such detectors are typically divided into angular sections, enabling precise reconstruction of scattering patterns and spin-dependent effects (Hejny 2016).

According to Dosch, spin and its discrete orientations represent a fundamental feature of nature (Dosch 2005: 47). Angular momentum therefore plays a central role in both classical and quantum physics. The smallest unit of angular momentum is expressed in terms of the reduced Planck constant ħ, and spin is described as a quantized angular momentum, often with half-integer values such as ½.

Particles with spin ½ can align themselves either parallel or antiparallel to a chosen axis, typically the z-axis of a coordinate system. These two possible states are commonly referred to as spin-up (+½) and spin-down (−½).

If a system consisting of such spin states is rotated by 180° around its axis, the physical configuration remains equivalent, although the order of the states may be reversed. This behavior illustrates the non-classical nature of spin and its distinction from ordinary mechanical rotation (Dosch 2005: 47).

This insight—that elementary particles possess a quantized spin with two preferred orientations—has formed the basis for a wide range of developments in modern physics.

3.1 The Spin of Protons and Neutrons

The two spin states, spin +½ and spin −½, apply to all particles of matter, including protons. Spin +½, or spin-up, is commonly associated with one orientation along a chosen axis, while spin −½, or spin-down, corresponds to the opposite orientation.

In modern physics, the spin of elementary particles is interpreted as intrinsic angular momentum, whereas the total angular momentum of composite particles includes both intrinsic spin and orbital contributions. Since nucleons are considered composite particles, their total angular momentum arises from the intrinsic angular momenta and orbital motions of their constituent subparticles. The direction of angular momentum vectors is conventionally described using the right-hand rule.

As early as 1933, Otto Stern and Immanuel Estermann succeeded in detecting the spin of the proton. By measuring the deflection of a hydrogen atom beam in a magnetic field, they were able to determine the magnetic moment of the proton and confirm its spin of ½ (Paetz 2020: 34). The magnetic moment of the neutron can be measured more easily when it is bound with a proton in a deuteron (Paetz 2020: 34–35).
Advances in experimental techniques have since enabled increasingly precise measurements of such properties. According to reports from the Max Planck Society, highly sensitive experiments have been developed to determine the magnetic moment of individual protons. These experiments employ devices

similar to a Penning trap, which makes it possible to isolate a single proton. Its presence can be detected through oscillatory motion that generates a measurable electrical signal (mpg 2014; Mooser et al. 2014).

The proton itself was first identified by Ernest Rutherford in 1919, while the neutron was discovered in 1932. These discoveries made it possible to develop models of nuclear structure. In this context, Werner Heisenberg proposed that protons and neutrons could be understood as two states of a single particle:

"(...) Proton und Neutron als ein und dasselbe Teilchen zu betrachten, das sich in zwei verschiedenen Zuständen befinden kann" (Mayer-Kuckuk 1992: 155).

"(...) to regard the proton and the neutron as one and the same particle that can exist in two different states.")

This concept is analogous to the two spin states of the electron, which can be described as spin ↑ (+½) and spin ↓ (−½). On this basis, Heisenberg introduced the concept of "isospin", which serves as an abstract quantum number distinguishing protons and neutrons. However, this quantity is not identical to the physical spin.

Mayer-Kuckuk describes this abstraction as follows: "Führt man aber für die Elektronen eine interne Variable s ein, die zu genau zwei Eigenzuständen ↑ und ↓ führt, so kann man alle Elektronen als gleich behandeln, aber sie haben nun einen internen Freiheitsgrad." (Mayer-Kuckuk 1992:155).

("If one introduces an internal variable s for electrons, leading to exactly two eigenstates ↑ and ↓, then all electrons can be treated as identical, but they possess an internal degree of freedom.")

In this framework, the two states of nucleons can be treated analogously to spin states. They are represented as two possible configurations within an abstract space. The concept of isospin was therefore mathematically formulated in analogy to angular momentum.
As Wolfgang Demtröder explains: "Die Ladungsunabhängigkeit der Kernkräfte zeigt, dass man Proton und Neutron hinsichtlich der starken Wechselwirkung als gleiche Teilchen (Nukleonen) ansehen kann, die sich nur durch ihre Ladung unterscheiden, d. h. nur hinsichtlich der Coulomb-Wechselwirkung verschieden sind"(Demtröder 2014: 373–374).

("The charge independence of nuclear forces shows that, with respect to the strong interaction, protons and neutrons can be regarded as identical particles (nucleons) that differ only in their charge, i.e., only with respect to the Coulomb interaction.")
In addition, Klaus-Peter Haupt emphasizes that isospin behaves mathematically in a manner analogous to real spin: "Der Isospin ist also ein Vektor in einem dreidimensionalen Isospinraum, so wie der Drehimpuls ein Vektor im dreidimensionalen Ortsraum ist" (Haupt 2022:1.00–1.10).

("Isospin is thus a vector in a three-dimensional isospin space, just as angular momentum is a vector in three-dimensional space.")

Thus, within the established theoretical framework, proton and neutron states are described using abstract quantum numbers that formally resemble spin. However, the physical interpretation of these quantities—and their possible relation to real rotational motion—remains a subject of ongoing discussion

3.2 Atomic Spin

Atoms can also possess a spin, which is described as total angular momentum or nuclear spin. This quantity arises from the contributions of protons and neutrons, which themselves carry spin and angular momentum. Nuclear spin is commonly denoted by *I* and includes both the intrinsic spins and the orbital angular momenta of the nucleons. It is associated with a magnetic moment (Mayer-Kuckuk 1992: 54).

Over the past decades, numerous researchers and research institutions have investigated the "spin of the proton." A central question has been the extent to which the spins of subparticles contribute to the total spin of the proton. These investigations led to what is commonly referred to as the "spin crisis" in particle physics, as experimental results did not match theoretical expectations.

More than three decades ago, Theo Mayer-Kuckuk described the internal structure of the proton as follows:

“Die Valenzquarks tragen also nur wenig zum Spin des Protons bei. Dieser wird im Wesentlichen durch Bahndrehimpulsbeiträge der Quark-Antiquark-Paare verursacht. Man spricht von der Spinanomalie des Protons. Das Proton hat also ein ziemlich kompliziertes Innenleben“ (Mayer-Kuckuk 1992: 183). (“The valence quarks thus contribute only slightly to the proton’s spin. This is essentially caused by orbital angular momentum contributions from quark–antiquark pairs. This is referred to as the proton spin anomaly. The proton therefore has a rather complex internal structure.”)

This problem has persisted for decades and has been the subject of extensive experimental and theoretical work. For example, reports in *Spektrum* emphasize that although the value of spin ½ is well known, the internal origin of this quantity remains unclear (Spektrum 2017). Experimental results at particle accelerators indicate that the intrinsic spins of the three valence quarks contribute only about one-third of the proton’s total spin:

“Experimente an Teilchenbeschleunigern haben gezeigt, dass die Eigendrehung der drei sogenannten Valenzquarks, aus denen das Proton nach vereinfachten Modellen besteht, nur ein Drittel des Spins beisteuert“ (Spektrum 2017).

("Experiments at particle accelerators have shown that the intrinsic spin of the three so-called valence quarks contributes only one-third of the proton's total spin.")

In many contexts, spin is defined more generally as a form of angular momentum. Reports from the Austrian Science Fund describe spin as a fundamental property of elementary particles, sometimes illustrated by analogy with rotating bodies such as spinning tops (FWF 2007; Bass 2007). In this interpretation, it is not the visible rotation itself that is decisive, but the associated angular momentum.

Similarly, the ETH Zurich defines spin as an intrinsic angular momentum with a magnitude of s = ½ ħ for nucleons. (ETH 2023: A4–A5). Within modern nuclear and particle physics, it is therefore generally accepted that protons and neutrons are composite particles whose total spin arises from multiple internal contributions.

Consequently, experimental investigations of the "spin structure" of the proton remain a major focus of international research. These studies aim to determine how much of the proton's spin is contributed by the spins and orbital motions of quarks and gluons. To address this question, large-scale international research programs have been conducted, involving major facilities such as CERN, DESY, Brookhaven National Laboratory, Jefferson Lab, and SLAC National Accelerator Laboratory (FWF 2007; openpr 2023). The results of these efforts have been summarized, among others, by Steven D. Bass in *The Spin Structure of the Proton* (Bass 2007).

More recently, new experimental findings have pointed to previously unknown forms of magnetic order. Studies on so-called “altermagnetism” indicate that atoms in certain materials can arrange themselves in patterns with alternating spin orientations (Krempask et al. 2024; Podbregar 2024). In such systems, antiparallel spins form ordered structures reminiscent of a checkerboard pattern.

This principle is also known from crystal structures such as sodium chloride, in which different atomic species alternate in a regular arrangement. In a comparable manner, atoms with spin-up and spin-down orientations can alternate within a lattice, contributing to the stability of the system.

These observations suggest that antiparallel spin configurations may play a fundamental role in stabilizing physical systems. A corresponding theoretical approach, proposing a checkerboard-like arrangement of nucleons, was already formulated in 2017 (Albert G., Albert H. 2017)

3.3 Angular Momentum and Torque

The concept of spin is closely related to the physical quantities of angular momentum and torque. On the archive page of the Austrian Science Fund, an article on proton spin describes an international research project from 2007 and emphasizes that spin is often associated with rotational motion:

“Many elementary particles rotate around their own axis like a spinning top. Unlike spinning tops, however, this torque—referred to as spin—has a fundamental influence on the properties of the elementary particle” (FWF 2007).

In this context, spin is not described merely as angular momentum, but is also associated with torque, highlighting its dynamical character. The discussion refers in particular to the spin contributions of quarks and gluons within the proton (FWF 2007; openpr 2023).

Angular momentum and torque are fundamental quantities in both classical and modern physics. In his standard work *What Is Spin?* Paetz describes spin as follows:

> *(“According to all the rules of quantum mechanics, spin has the character of angular momentum” (Paetz 2020: 23).)*

He further explains that, within quantum theory, spin must be understood as a quantized form of angular momentum, even when it does not correspond to classical mechanical rotation. In composite particles, total spin arises from a combination of intrinsic spin and orbital angular momentum, whereas in elementary particles it is attributed solely to intrinsic angular momentum (Paetz 2020: 50).

In classical mechanics, a distinction is made between translational motion and rotational motion. A force acting on a free body causes linear acceleration, while a force applied to a body fixed at a point results in rotational motion. This rotational effect is described by torque.

According to the University of Göttingen, a force acting on a body that is constrained at a fixed point produces a change in angular velocity, which is determined by the applied torque. The torque is defined as the product of the force and the perpendicular displacement vector from the point of rotation (University of Göttingen 2006–2014).

The direction of the torque vector can be determined using the right-hand rule. In this formulation, the torque vector points in the same direction as the angular momentum vector. Thus, torque and angular momentum are closely related quantities describing rotational motion.

Although translational forces and rotational torques differ in their physical manifestation, they are formally analogous. Many physical motions involve a combination of both types of movement, and similar mathematical methods can be used to describe them (Physki 2023).

These relationships show that rotational motion, angular momentum, and torque are fundamentally interconnected. This raises the question of whether the spin of particles—traditionally interpreted as an intrinsic quantum property—may also be related more directly to real rotational processes and the forces associated with them.

Chapter 4 - Force Effects

Forces become evident through their effects on physical objects, as bodies undergo motion or deformation. In this process, there is always a distinguishable "before" and "after" state of the affected matter, while the force itself remains immaterial and cannot be observed directly.

Because forces are not directly accessible, their description has often relied on indirect concepts and terminology. In physics, the term "force" is frequently used interchangeably with "interaction." Accordingly, the weak and strong interactions have historically also been referred to as the "weak force" and the "strong force."

Long before the development of modern physics, Isaac Newton established a precise definition of force in the 17th century through his laws of motion. In doing so, he laid the foundation of classical mechanics, which remains fundamental to this day. According to Newton, a force manifests itself through a change in the motion or direction of a body. His third law, commonly expressed as *actio = reactio*, describes the mutual interaction of forces and is also known as the principle of action and reaction ("Newton's Laws" 2024).

Immanuel Kant later adopted this framework and interpreted attraction and repulsion as the fundamental forces of the universe in his early cosmological writings (Kant 1755; 1925: 19).

In the preceding chapters, it has been shown that modern physics explains the fundamental forces through interaction theories. Matter is described in terms of fermions and bosons. While fermions are particles with spin ½, bosons are assigned integer spin values (0, 1, or 2) and are interpreted as exchange particles. According to this framework, forces arise through the exchange of such particles between fermions.

In their standard work *Mechanics and Thermodynamics*, Roth and Stahl illustrate the concept of force using classical examples. They describe the principle of momentum transfer as follows:

"One can make this recoil visible by standing on a moving surface, for example in a boat on a lake. After throwing an object, the boat moves backward. The same effect occurs when catching a heavy object: the force required to stop it results in a reaction that moves the boat in the opposite direction" (Roth and Stahl 2016: 1952f.).

Such illustrative examples are frequently used in the literature to explain the observable effects of forces. However, they primarily describe the outcomes of interactions rather than their underlying causes.

This distinction raises a fundamental question: while the effects of forces can be observed and measured, what their underlying origin is. If forces themselves remain immaterial and are inferred only through their effects, it becomes necessary to examine whether a more fundamental physical mechanism may exist that gives rise to these interactions.

Chapter 5 – The Fundamental Spin Force

This chapter presents the concept of the fundamental spin force. According to this approach, it is proposed that there are not four separate fundamental forces, but a single underlying interaction: the fundamental spin force. The following sections outline the basic ideas of this model. The two fundamental interactions, attraction and repulsion, are interpreted as arising exclusively from the spin properties of particles and objects. In this context, a connection is established between the laws of Newton and Coulomb and the concept of spin discovered in the 20th century. Within this framework, spin is interpreted as a real physical rotation.

In this context, "right-handed" and "left-handed" refer to the orientation defined by the right-hand rule and do not correspond to clockwise or counterclockwise rotation. Thus, the Earth can be described as a right-handed (prograde) rotating body.

Within this model, all right-handed "free protons" are assumed to share the same axis orientation and to form the basic constituents of matter. Accordingly, the universe is interpreted as anisotropic, meaning that a preferred orientation ("up" and "down") can be defined. This can be illustrated by a spin vector that points upward for right-handed protons and downward for neutrons. According to the right-hand rule, spin-up corresponds to right-handed rotation, whereas spin-down corresponds to left-handed rotation. Nucleons with parallel spin states repel each other, while nucleons with antiparallel spin states attract each other. This principle is proposed as the basis for a checkerboard-like planar structure of atoms.

The rotation of astronomical objects such as planets and stars is described as prograde or retrograde, corresponding to right- or left-handed rotation. As early as the 16th century, Nicolaus Copernicus demonstrated that the Sun is at the center of the solar system and that planets rotate about their own axes. The arrangement of such systems in concentric structures is primarily determined by differences in size and mass, as illustrated by the solar system. If planets and the Sun were equal in size and mass, they would—by analogy with composite atomic structures—form a checkerboard-like arrangement with alternating rotational orientations. In systems with unequal masses, smaller bodies can compensate for the influence of larger ones through opposing rotational states, consistent with classical lever principles.

It follows that all particles and objects in the universe—or their constituents—are in motion and exhibit rotational behavior. While astronomical bodies follow observable trajectories, particles may exhibit oscillatory, vibrational, and wave-like behavior depending on their environment.

Every rotation, or real spin, represents accelerated motion associated with a force. This force can be described in terms of torque, which can act in two opposite directions. While angular momentum characterizes the rotational state, the sign of the torque determines the direction of rotation. The force associated with the rotating building blocks of matter is referred to here as the fundamental spin force, as it is not reduced further within this model. Rapidly rotating bodies with similar torque orientations repel each other, whereas bodies with opposite torque orientations attract each other.

By analogy with Coulomb's law, positive and negative electric charges can be reinterpreted in terms of opposite torque orientations without altering the resulting attractive and repulsive behavior. In this sense, the fundamental spin force provides a unified description of both rotational motion and the interactions of attraction and repulsion. The observed duality of these interactions is thus associated with the duality of opposite rotational states: bodies with the same rotational orientation repel each other, while those with opposite orientations attract each other.

What is a Proton?

Within the framework of the proposed checkerboard-like planar atomic structure, the proton is interpreted as a rapidly rotating elementary unit. In this model, the proton is not assumed to be perfectly spherical, but rather to possess a flattened, disc-like geometry, comparable to a spinning top or a discus.

A central assumption of this approach is that free protons share a common orientation of their rotational axis. This alignment is considered stable as long as no external interaction disturbs the system. In this sense, the proton is treated as a fundamental building block of matter.

The diversity of observed physical phenomena is, within this framework, attributed to different states and configurations of protons and neutrons. Without neutrons, the formation of atoms beyond hydrogen would not be possible.

The model further assumes that, under certain conditions—such as proton collisions—the rotational axis of a proton may invert by approximately 180°, resulting in a transition to a neutron state. Conversely, when such a neutron is no longer bound within an atomic structure, it is proposed to return to the proton state.

In addition, it is assumed that oscillatory motions of protons and neutrons may give rise to magnetic effects.

What is a Elektron?

According to classical physics, electrons exist both as free particles and as particles bound within atoms. They are assumed to carry a negative electric charge and to play a central role in electrical phenomena. The flow of electric current is commonly explained by the motion of electrons in conductive materials. Within atoms, electrons are also regarded as responsible for chemical bonding.

However, within the framework of quantum mechanics, electrons are not described as well-defined particles in the classical sense. According to the wave–particle duality, they exhibit both particle-like and wave-like properties and can only be described within the atomic shell in terms of probability distributions.

Within the framework of the present model, this interpretation is reconsidered. According to the theory of a checkerboard-like planar atomic structure, electrons do not constitute an independent class of particles. Instead, they are interpreted as free protons in a different (oscillatory) physical state.

In this view, the state of free protons differs significantly from that of protons bound within atomic structures. It is assumed that free protons undergo oscillatory motion as a result of the Earth's rotation and therefore behave differently from bound protons, whose motion is strongly constrained. This difference in state is proposed as a possible explanation for the phenomena commonly attributed to electrons.

What exactly is the "weak interaction"?

In contemporary physics, the weak interaction is understood as the mechanism responsible for processes such as the conversion of protons into neutrons, for example in nuclear fusion within stars, as well as the transformation of neutrons into protons during radioactive decay. Within the Standard Model, these processes are described in terms of interactions mediated by gauge bosons.

Within the framework of the present model, an alternative interpretation is proposed. The transformation between protons and neutrons is not attributed to the exchange of bosons or to a restructuring of subatomic constituents, but is instead associated with a change in the rotational state of the nucleon.

More specifically, the transition is interpreted as a reversal of the rotational axis by approximately 180°, referred to here as a "spin flip." In this view, processes such as nuclear fusion involve collisions between protons, during which the rotational axis of one or more particles may invert. This inversion leads to a transition from one rotational state to another, corresponding to what is conventionally described as the transformation from proton to neutron.

In this model, neutrons are considered stable only within bound atomic systems. Their stability is associated with their integration into a structured arrangement of nucleons. When a neutron exists in isolation, the model assumes that its rotational configuration is not stable and tends to return to the preferred proton orientation.

This process can be described as a gradual realignment of the rotational axis. Over time, the neutron transitions back into the proton state, accompanied by an oscillatory adjustment of its rotational configuration.

Accordingly, the transformation between proton and neutron states is interpreted as a dynamical process governed by rotational properties. Collisions between particles can induce such transitions by influencing the orientation of their rotational axes. Conversely, the absence of stabilizing interactions allows the system to revert to its original state.

It is noteworthy that controlled reversals of spin states—commonly referred to as "spin flips"—are experimentally realizable in modern physics. Such processes are investigated in the field of spintronics, where external fields are used to manipulate the spin states of particles and materials.

In this sense, the distinction between the experimentally observed spin flip and the transformations described here may be understood as a difference in scale and context, rather than as fundamentally different processes. The underlying mechanism is interpreted as a change in intrinsic angular momentum, governed by the fundamental force of spin.

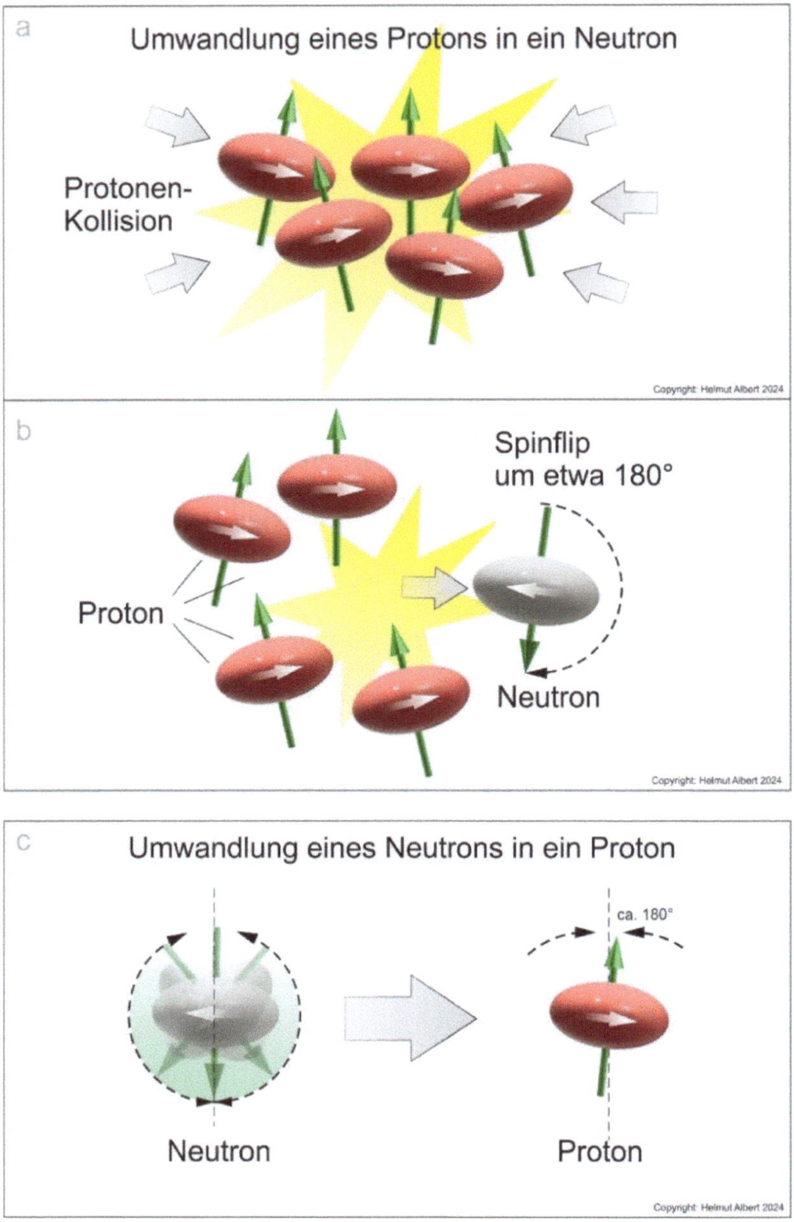

Fig. 1a. Proton collision. Fig. 1b. Proton conversion.

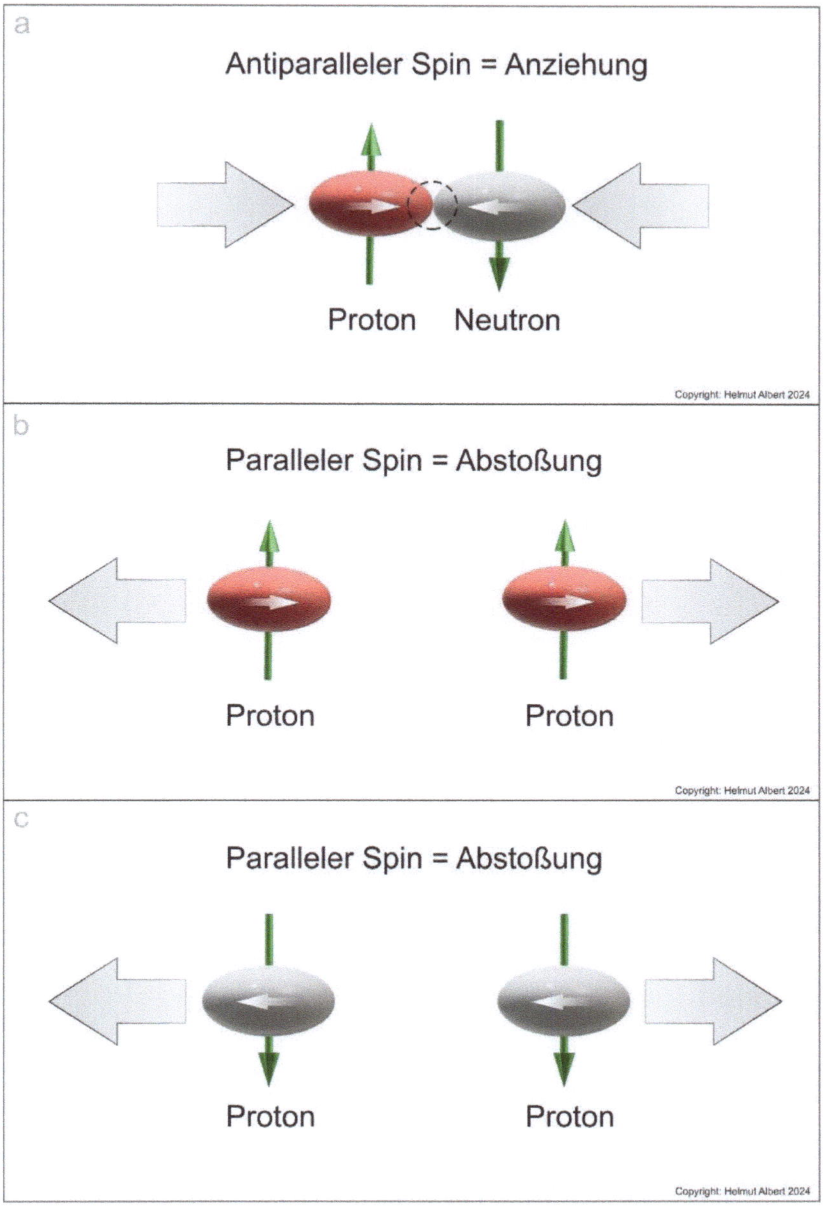

Fig. 1c. Neutron conversion. Fig. 2a. Proton-neutron attraction. Fig. 2b. Proton repulsion. Fig. 2c. Neutron repulsion

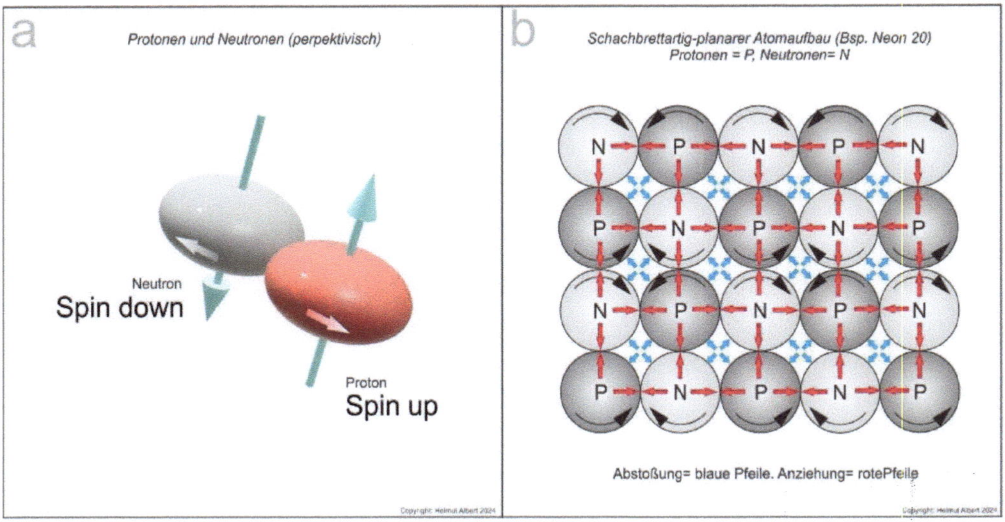

Fig. 3a. Spin-up, spin-down, nucleons. Fig. 3b. Checkerboard-like atomic structure

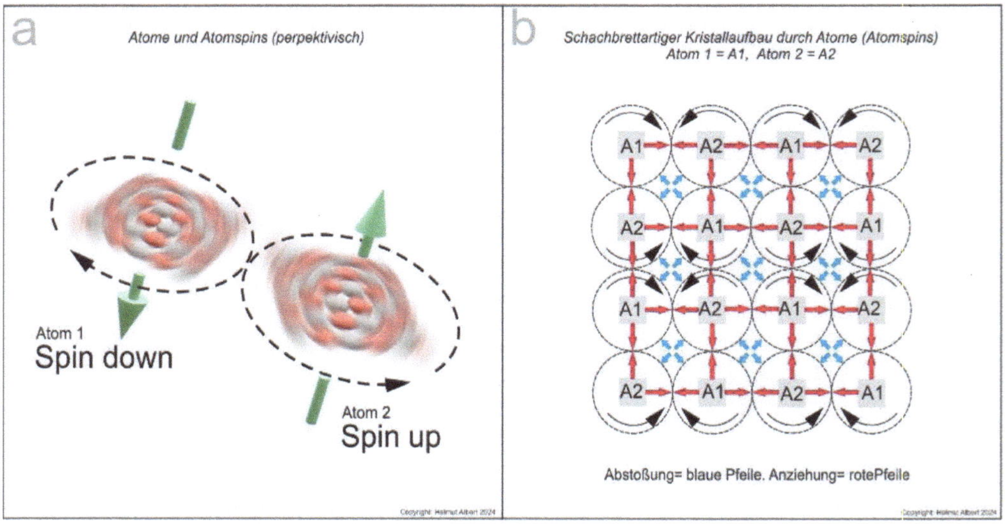

Fig. 4a. Spin-up, spin-down, atoms. Fig. 4b. Checkerboard-like crystal structure

What is the "electromagnetic interaction" really?

Within modern physics, the electromagnetic interaction is described with high precision in terms of its observable effects. However, the underlying nature of electric charge itself remains conceptually abstract. While the Standard Model provides a highly successful mathematical framework, it does not fully explain what electric charge fundamentally *is*, but rather how it behaves.

Since the 18th century, Charles-Augustin de Coulomb formulated the quantitative law governing electric forces. Coulomb showed that two types of charge—designated as positive and negative—give rise to attraction and repulsion. Charges of the same sign repel each other, while charges of opposite signs attract each other.

In the framework of the present model, an alternative interpretation is proposed. Instead of treating electric charge as a fundamental and irreducible property, the observed effects of attraction and repulsion can be related to the rotational properties of particles.

In this interpretation, the two types of electric charge correspond to two opposite rotational states. Positive and negative charges are thus associated with opposing orientations of rotational motion, or, equivalently, with opposite signs of torque.

If this correspondence is assumed, the behavior described by Coulomb's law can be reproduced: systems with identical rotational orientation exhibit repulsive behavior, while systems with opposite rotational orientation exhibit attraction.

This suggests that the electromagnetic interaction may not require an independent fundamental origin, but could instead be understood as a manifestation of more general rotational dynamics. In this sense, the duality of positive and negative charge reflects the duality of rotational states.

Accordingly, the fundamental force of spin provides a unified basis for both rotational motion and the interactions traditionally attributed to electric charge.

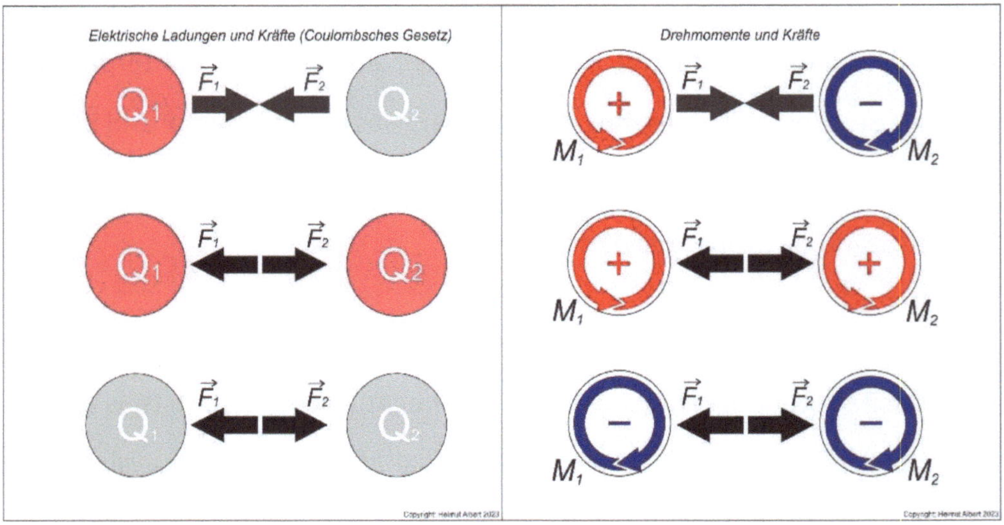

Fig. 5a Coulomb's law. Fig. 5b Positive and negative torque.

What is "gravity" really?

Gravity occupies a special position among the known interactions, as it is observed predominantly as an attractive effect. In classical physics, Isaac Newton described gravity as a force acting between masses. In modern physics, Albert Einstein reinterpreted gravity as a manifestation of the curvature of spacetime.

Both approaches provide highly successful descriptions of gravitational phenomena. However, they primarily characterize how gravity acts, rather than offering a direct mechanical interpretation of its underlying physical origin.

Within the framework of the present model, gravity is interpreted as a directional interaction arising from rotational dynamics. In this view, gravitational effects are associated with the collective behavior of rotating particles within a body.

When a macroscopic object such as the Earth rotates, its constituent particles exhibit different dynamical conditions depending on their position. Particles near the center experience minimal motion relative to the rotational axis, while particles at larger radii are subject to higher velocities. This distribution leads to a gradient of rotational motion within the body.

In this model, such a gradient gives rise to an effective inward-directed tension. The outer regions of the system are dynamically coupled to the inner regions, resulting in a tendency toward the center. This effect is interpreted as the physical origin of gravitational attraction.

In addition, it is assumed that free particles within and around the rotating body adjust their rotational orientation relative to the global axis of rotation. Small deviations from this preferred orientation may lead to oscillatory motion. The collective oscillations of such particles can contribute to large-scale electromagnetic effects, such as planetary magnetic fields (Albert G.; Albert H. 2017; mpg 2014).

Particles that are bound within stable atomic structures are less able to compensate for these oscillations. As a result, they experience a net inward-directed effect, which manifests macroscopically as gravitational attraction.

To illustrate this principle, it is useful to consider analogies from classical electrodynamics, such as the Faraday disk. In such systems, rotational motion can generate measurable electrical potentials between the center and the outer regions. This demonstrates that rotation can give rise to field-like effects.

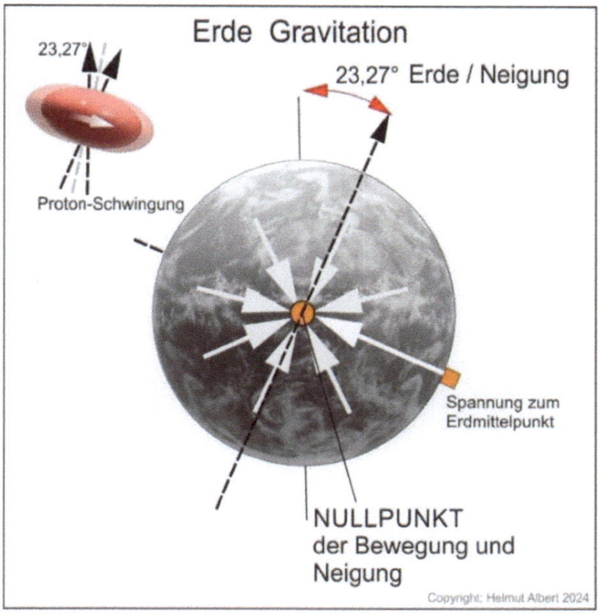

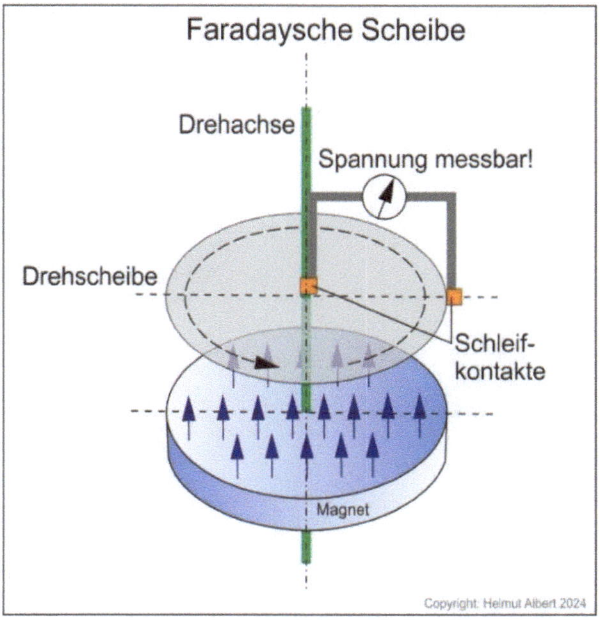

Fig. 6a Earth's axial tilt; Fig. 6b Faraday disk

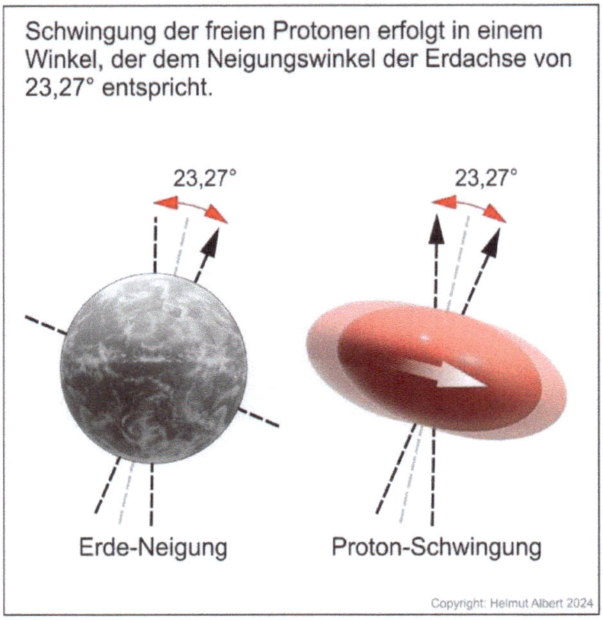

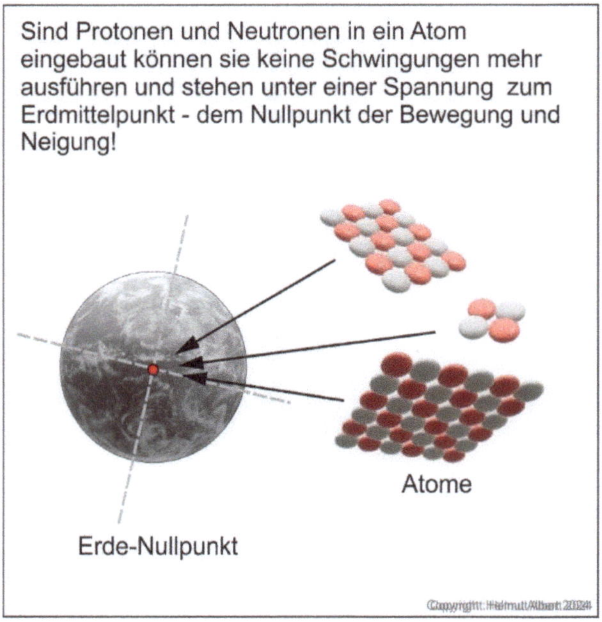

Fig. 7a Tilt and oscillation. Fig. 7b Gravitational attraction

In this sense, the fundamental force of spin is proposed to act not only at the level of elementary particles, but also in macroscopic and astronomical systems.

From this perspective, the large-scale structure of planetary systems and galaxies may also be influenced by rotational dynamics. While gravitational models successfully describe many observed phenomena, the present approach suggests that attraction and repulsion may arise from a more general principle related to spin and rotational motion.

The arrangement of astronomical bodies can therefore be interpreted as the result of their angular momenta and rotational states. In contrast to atomic systems—where comparable constituents may form alternating structures—astronomical systems consist of bodies with vastly different masses. As a result, they form concentric configurations around dominant central bodies rather than checkerboard-like arrangements.

Experimental Setup for Demonstrating Attraction and Repulsion

One approach to investigating the effects of rotational motion on interaction behavior is the experimental arrangement shown in Fig. 8, which consists of two rapidly rotating bodies suspended in close proximity.

In configuration (A), two bodies rotating in the same direction are suspended from a flexible support, such as a rope or cable, allowing limited lateral movement. Within the framework of the present model, it is expected that bodies with identical rotational orientation exhibit a tendency to move apart, corresponding to a repulsive interaction.

In configuration (B), two bodies rotating in opposite directions are arranged in an analogous manner. In this case, the model predicts a tendency for the bodies to move toward one another, corresponding to an attractive interaction.

This experimental setup is intended to illustrate how differences in rotational orientation may influence interaction behavior. It provides a conceptual and potentially observable demonstration of the proposed relationship between rotational dynamics and the phenomena of attraction and repulsion.

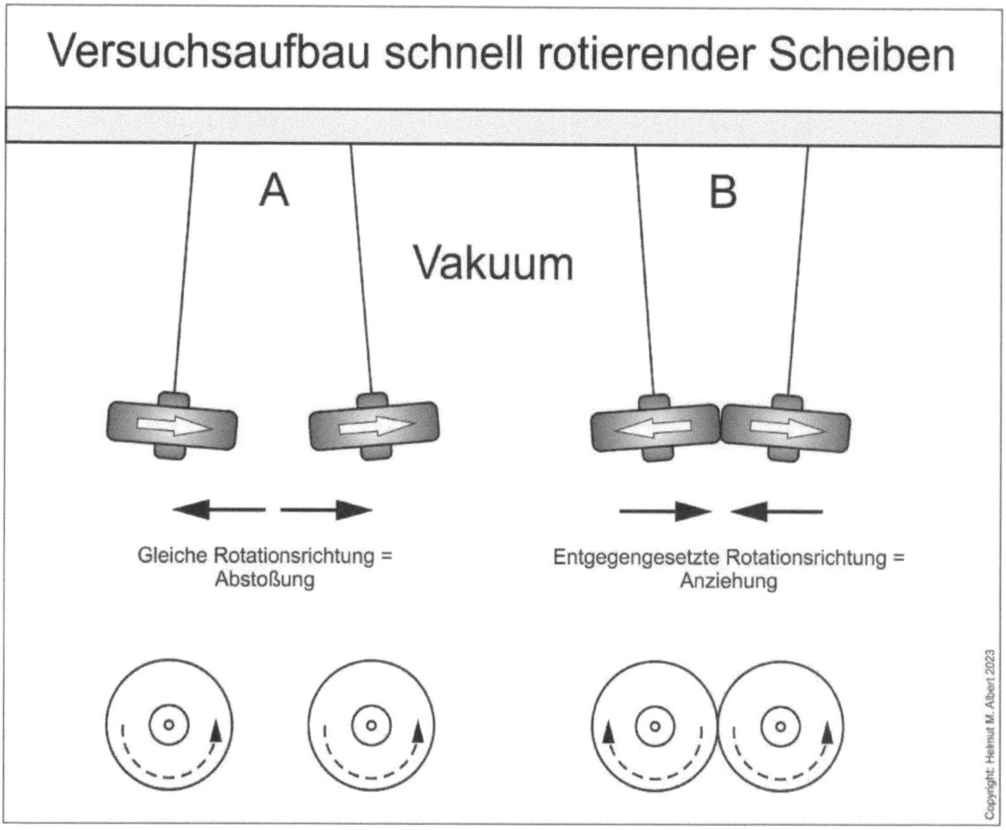

Fig. 8. Experimental setup: clockwise and counterclockwise rotating disks

Summary

This work investigates whether fundamental interactions can be explained in a mechanically interpretable way. In contrast to the Standard Model, which describes interactions through the exchange of force-carrying particles, spin is interpreted as a real physical rotation of particles and objects.

It is shown that attraction and repulsion arise from the relative orientation of torques or spin states: aligned states lead to repulsion, while opposite states lead to attraction. On this basis, the fundamental spin force is proposed as a unifying principle governing both microscopic and macroscopic structures.

This approach offers an alternative conceptual framework for describing fundamental interactions and connects the rotational properties of particles with large-scale structures of the universe.

Bibliography

Acharya Shreyasi; ALICE Collaboration (2020). Unveiling the strong interaction among hadrons at the LHC. Nature, 9. Dezember 2020 – https://doi.org/10.1038/s41586-020-3001-6

Albert, Helmut (Hg.) (2023). 2d-Atome; Kernformen und Kernstruktur. Verlag Helmut Albert, Freiburg. BoD –Druck, Norderstedt.

Albert, Helmut (Hg.) (2020a). The Nucleon Spin as Cause oft the Strong Interaction; Verlag Helmut Albert, Freiburg.Vertrieb: Epubli, Berlin. Veröffentlichung (ePUB) ISBN: 978-3-753139-15-9

Albert, Helmut (Hg.) (2020b). Der Nukleonenspin als Ursache der Starken Wechselwirkung; Verlag Helmut Albert, Freiburg.Vertrieb: Epubli Druck, Berlin. (Hardcover) ISBN: 978-3-753137-33-9

Albert, Helmut (Hg.) (2019). Die Spin-Kernkraft; Planares Atommodell; Verlag Helmut Albert, Freiburg. Epubli –Druck, Berlin.

Albert, Helmut (Hg.) Albert Georg (2017). Atommodell mit schachbrettartiger Struktur, Verlag Helmut Albert, Freiburg. Epubli – Druck, Berlin.

Bass Steven D. (2007).The Spin Structure of the Proton. By Steven D. Bass, Publisher World Scientific, ISBN 978-981-270-9479.

Bleck-Neuhaus Jörn (2013). Elementare Teilchen: Von den Atomen über das Standard-Modell bis zum Higgs-Boson (Springer-Lehrbuch) (German Edition) Kindle-Version.

Bergische Universität Wuppertal (2023a). Es gibt nur vier Kräfte. http://www.atlas.uni-wuppertal.de/oeffentlichkeit/4Kraefte.html [Abgerufen: 27.11.2023]

Bergische Universität Wuppertal (2023b).Das Standard-Modell. http://www.atlas.uni-wuppertal.de/oeffentlichkeit/Parameter.html [Abgerufen: 27.11.2023]

Brandenburgische Technische Universität Cottbus. (PDF) M10. Physikalisches Praktikum. Grundgesetze der Rotation.

Camejo Sylvia Arroyo (2015). Skurile Quantenwelt, 4. Auflage 2015. Ersterscheinung 2007. Fischer Taschenbuch. Frankfurt am Main.

„Coulombsches Gesetz"(2024). In: Wikipedia – Die freie Enzyklopädie. Bearbeitungsstand: 26. November 2023, 09:26 UTC. URL: https://de.wikipedia.org/w/index.php?title=Coulombsches_Gesetz&oldid=239473705 (Abgerufen: 2. Mai 2024, 09:01 UTC)

Demtröder Wolfgang (2014). Springer-Lehrbuch, Experimentalphysik 4, 4. Aufl. 2014, Kern-, Teilchen- und Astrophysik, DOI: 10.1007/978-3-642-21476-9, Springer-Verlag Berlin Heidelberg2014. Kindle-Version.

DESY (2023a) o.V.: Lexikon. Desy´s KworkQuark. Deutsches Elektronen-Synchrotron. DESY http://kworkquark.desy.de/lexikon/lexikon.virtuelleteilchen/1/index.html [Abgerufen: 26.11.2023]

DESY (2023b) o.V.: Lexikon. Desy´s KworkQuark. Deutsches Elektronen-Synchrotron. http://kworkquark.desy.de/lexikon/lexikon.gluon/1/index.html) [Abgerufen: 27.11.2023]

Dosch Hans Günter (2005). Jenseits der Nanowelt. Leptonen, Quarks und Eichbosonen. Springer; 2005. Edition. Springer-Verlag Berlin, Heidelberg, New York. Kindle-Version

Escher U., Schwab A. (2010). PDF. Fachrichtung Physik Physikalisches Grundpraktikum. TU Dresden. 2 Allgemeine Grundlagen 2.1 Bewegung des starren Körpers.

ETH Zürich (2023). Aufbau und Eigenschaften der Atomkerne. PDF. https://www.analytik.ethz.ch › Nukleonik

FWF (2007) https://Protonenspin-was ist das? fwf.ac.at/de/wissenschaft-konkret/projektvorstellungen-archiv/2007/pv200712 .Österreichischer Wissenschaftsfond [Abgerufen26.10.2023]

Goethe Universität Frankfurt (2023).o.V. https://aktuelles.uni-frankfurt.de/forschung/physik-neue-methode-fuer-praezisionsstudien-zur-starken-wechselwirkung/ [Abgerufen: 27.11.2023]

Georg-August-Universität Göttingen (2023). https://lp.uni-goettingen.de/get/text/3870 [Abgerufen: 27.11.2023]

„Gravitation". In: Wikipedia – Die freie Enzyklopädie. Bearbeitungsstand: 16. Januar 2024, 22:10 UTC. URL:https://de.wikipedia.org/w/index.php?title=Gravitation&oldid=241248434 (Abgerufen: 29. März 2024, 10:32 UTC)

Haupt Klaus-Peter (2022a). Was ist die Kernkraft? Warum gibt es genau 8 Gluonen? | KP Haupt - Elementarteilchen Teil VIII. Vortragsreihe Elementarteilchen SFN-Kassel. [Video] Youtube. https://www.youtube.com/watch?v=Aw2LGfuiGH0&list= PLJWxm2yBpirgN3vvoYf2_O1J-rs3JlPnw&index=8 [Abgerufen: 26.11.2023]

Haupt Klaus-Peter (2022b). Spin-Isospin? Spin, Isospin, schwacher Isospin? | KP Haupt - Elementarteilchen Teil II. Vortragsreihe Elementarteilchen SFN-Kassel. [Video] Youtube. https://www.youtube.com/watch?v=aCh5qi1inlE&t=4137s [Abgerufen: 27.11.2023]

Harms Volker (2000). Physik, für Mediziner und Pharmazeuten. Harms Verlag – Lindhöft. Herstellung: AZ Druck und Datentechnik Gmbh, Kempten.

Kant Immanuel (1925). Allgemeine Naturgeschichte und Theorie des Himmels. Die Kant-Laplace'sche Theorie. Alfred Kröner Verlag, Leipzig. Heinrich Schmidt (Hg.) (Jena).

Krempask, J., Mejkal, L., D'Souza, S.W. *et al.* (2024). Altermagnetisches Heben von Kramers Spin-Denereracy. *Natur* 626, 517–522 (2024). https://doi.org/10.1038/s41586-023-06907-7

Mayer-Kuckuk Theo (1992). Kernphysik. Eine Einführung. 5. überarbeitete und erg. Auflage. Stuttgart Teubner. Teubner Studienbücher: Physik

Meschede, Dieter (2020). Gerthsen Physik (Springer-Lehrbuch) Springer Berlin Heidelberg. Kindle-Version.

„Meson"(2023). In: Wikipedia – Die freie Enzyklopädie. Bearbeitungsstand: 30. Oktober 2023, 12:27 UTC. URL: https://de.wikipedia.org/w/index.php?title=Meson&oldid=238634445 [Abgerufen: 27. November 2023, 16:23 UTC]

Mooser Andreas, Stefan Ulmer, Klaus Blaum, Kurt Franke, Holger Kracke, Clemens Leiteritz, Wolfgang Quint, Crícia de Carvalho Rodegheri, Christian Smorra & Jochen Walz (2014). Direct high-precision measurement of the magnetic moment of the proton Nature 29 May 2014; DOI: 10.1038/nature13388

MPG (2014). Am Kern des Antimaterie-Rätsels. Max-Planck-Gesellschaft. https://www.mpg.de/8232171/proton_magnetisches_moment [Abgerufen: 27.11.2023]

„Newtonsche Gesetze"(2024). In: Wikipedia – Die freie Enzyklopädie. Bearbeitungsstand: 21. März 2024, 16:24 UTC. URL:

https://de.wikipedia.org/w/index.php?title=Newtonsche_Gesetze&oldid=243318783 (Abgerufen: 3. April 2024, 11:22 UTC)

Openpr (2023).https://www.openpr.de/news/177970/Protonen-alles-dreht-sich-um-den-Spin.html [Abgerufen: 27.11.2023]

Physki (2023). Drehungen und Drehimpulssatz. https://www2.physki.de/PhysKi/index.php/Drehungen_und_Drehimpulssatz#Zentralkraft [Abgerufen: 27.11.2023]

Podbregar Nadja (2020). Unter extremem Druck verändert die Grundkraft ihre Wirkung auf die Kernbausteine. Scinexx. Das Wissensmagazin. [online] https://www.scinexx.de/news/technik/starke-kernkraft-ueberrascht-physiker/ [Abgerufen: 10.11.2020]

Podbregar Nadja (2024). Dritte Form des Magnetismus bestätigt. Scinexx. 19. Februar 2024. https://www.scinexx.de/news/physik/dritte-form-des-magnetismus-bestaetigt/ (Abgerufen 13.03.2024)

„Proton"(2023). In: Wikipedia – Die freie Enzyklopädie. Bearbeitungsstand: 9. November 2023, 04:40 UTC. URL: https://de.wikipedia.org/w/index.php?title=Proton&oldid=238941326 [Abgerufen: 28. November 2023, 15:08 UTC]

Roth Stefan, Stahl Achim (2020). Mechanik und Wärmelehre (German Edition). Springer Berlin Heidelberg. Kindle-Version.

Schmidt. A.; Pybus, J. R. ; Weiss, R. et al. (2020). Probing the core of the strong nuclear interaction. Artikel. Publiziert: 26.02.2020. in: Nature, vol. 578, pages 540–544 [online] https://doi.org/10.1038/s41586-020-2021-6 [Abgerufen:27.11.2023]

Spektrum (2017).o.V. Teilchenphysik: Spin des Protons https://www.spektrum.de/magazin/spin-des-protons/1446179[Abgerufen:27.11.2023]

Yukawa-Potential (2023): Lexikon der Astronomie. https://www.spektrum.de/lexikon/astronomie/yukawa-potential/541[Abgerufen:27.11.2023]

List of Figures

Biography

Helmut M. Albert

Born in 1960 in Speyer, Germany

Lives and works in Freiburg, Germany

1978–1981 Training in the art trade

Evening Academy Mannheim: life drawing and art history

1981–1987 Studied sculpture at the Academy of Fine Arts, Nuremberg

(Works based on the human figure, material studies, and work with steel, wood, plaster, and rubber)

1987–2006 Concrete art

Works based on material studies

Engagement with fundamental physics

Awards (selection):

2013 Artur Fischer Inventor Award of the State of Baden-Württemberg

1995 One-year scholarship for fine arts, Offenburg

Various art awards